The Six Sigma Design for Reliability

Volume II – The Improvement Process

Authored by

Darrin J. Wikoff

Authored with

Shon E. Isenhour

I0504415

Institute at Patriots Point • 40 Patriots Point Road
Mt. Pleasant, SC 29464
Phone 843.375.8222 • www.theinstituteatpatriotspoint.com

TABLE OF CONTENTS

PREFACE

The Six Sigma Design for Reliability series is a continuation of the *Leadership for Asset Management Excellence* book series. In collaboration with industry and academic leaders, this book is intended to be used as a resource for designing, administering and evaluating strategic asset management plans. This book is not accredited by the International Standards Organization, American National Standards Institute or any other governing standards body associated with ISO 55001. The views and perspectives expressed within this resource are those of the author based on his collective experience as a member of asset management councils and associations, and as a business and community leader in asset management.

In this second volume of the *Six Sigma Design for Reliability*, we will explore the foundational principles associated with a Six Sigma improvement project. We will outline the Define, Measure, Analyze, Improve and Control (DMAIC) model for managing your improvement project, and provide a detailed description of techniques used to derive the improvement hypothesis – conclusions drawn based on experimentation – that will help you explain why the current condition is incapable of meeting stakeholder expectations. Through measurement methods like "Maintainability" and "Reliability" we will discover the improvement priorities. Finally, this volume will discuss when and how to use "Root Cause Analysis", "Failure Modes and Effects Analysis" and "Graphical Analysis" techniques to recognize the source of suboptimal performance in order to begin the improvement phase of your project and drive a better future value for your business.

About the Author

Darrin J. Wikoff specializes in Organizational Change Management, Lean Manufacturing, Business Process Re-Engineering and Reliability Engineering. Since 2001, Darrin has continued to train, coach and mentor industrial leaders through the rigorous process of implementing and managing improvement initiatives in support of Lean Manufacturing. His ability to translate Lean and reliability engineering principles into real-world application has enabled more than 60 organizations to improve safety and environmental performance, increase production capacity and reduce operating costs. Darrin has also authored *"Centered On Excellence"* published by MRO-Zone in 2012 and the 7th edition of the *"Maintenance Engineering Handbook"* published by McGraw-Hill in 2008.

Shon Isenhour is an engineering graduate of North Carolina State University, the past National Chairman of the Society of Maintenance and Reliability Professionals (SMRP) and the current Chairman for the South Carolina chapter of the American Society for Training and Development (ATD, formerly ASTD). Since 2002, he has led improvement initiatives that have enabled his clients to succeed over their competitors in a changing global economy. As a Certified Maintenance & Reliability Professional (CMRP), Shon demonstrates superior technical subject matter knowledge and his ability as an experienced change management practitioner within such industries as primary metals, mining, pharmaceuticals, petrochemical, chemical processing and paper.

THE IMPROVEMENT PROCESS

In Volume 1 of the *Six Sigma Design for Reliability* we learned the practices associated with identifying strategic objectives for improvement projects in support of business opportunities to increase revenue, reduce maintenance costs, and reduce the company's asset base in order to generate a higher overall Return On Assets (ROA).

We also discovered how to quantify the improvement business case by converting performance measures, such as Availability, Overall Equipment Effectiveness and workflow Backlog, into common financial ratios that illustrate how improvements will impact the profitability and equity of the business.

Now it is time to discuss how we will manage the improvement process within the *Six Sigma Design for Reliability*. The model we will use, similar to other six sigma improvement efforts, is the "DMAIC" model. DMAIC is an acronym that represents the sequential steps of the improvement process that will ensure a measureable and meaningful return on the chartered improvement project and investment.

Figure 1 - iBL DMAIC Model

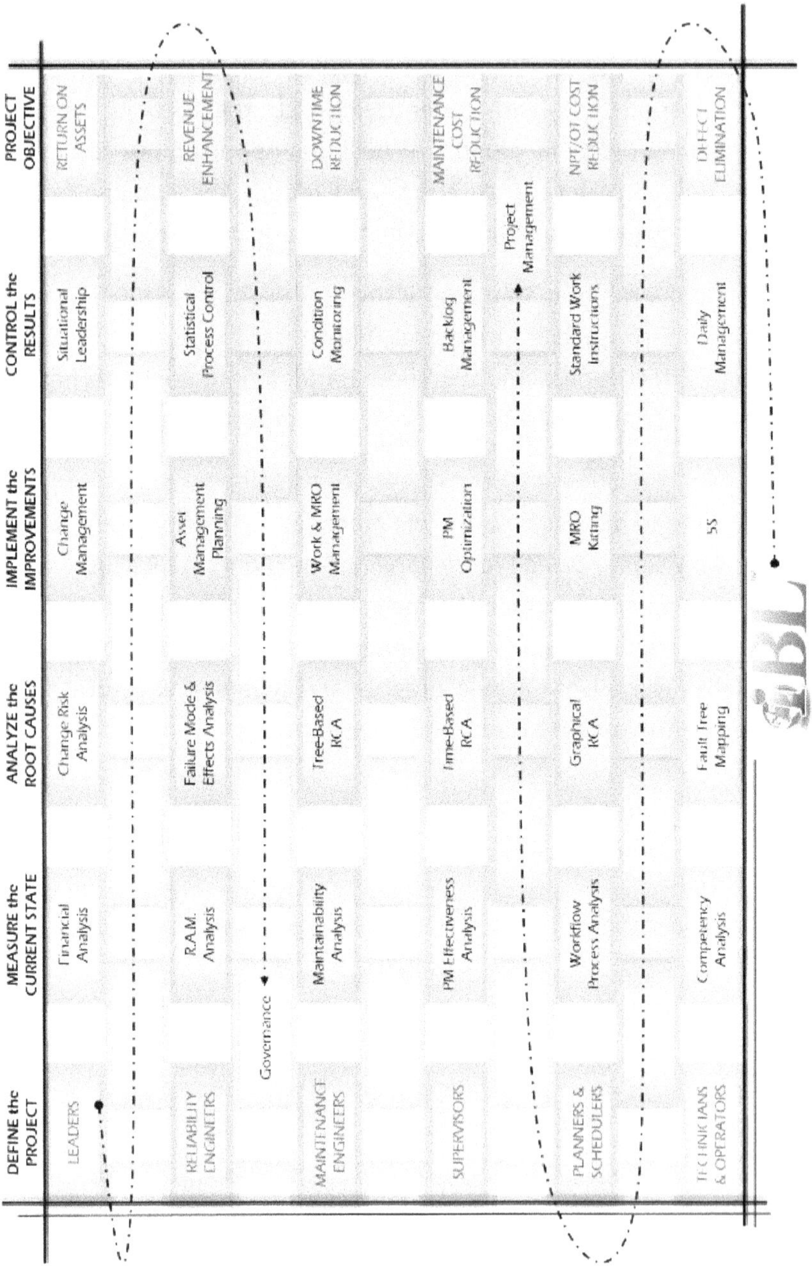

	DEFINE the PROJECT	MEASURE the CURRENT STATE	ANALYZE the ROOT CAUSES	IMPLEMENT the IMPROVEMENTS	CONTROL the RESULTS	PROJECT OBJECTIVE
LEADERS		Financial Analysis	Change Risk Analysis	Change Management	Situational Leadership	RETURN ON ASSETS
RELIABILITY ENGINEERS		R.A.M. Analysis	Failure Mode & Effects Analysis	Asset Management Planning	Statistical Process Control	REVENUE ENHANCEMENT
MAINTENANCE ENGINEERS		Maintainability Analysis	Tree-Based RCA	Work & MRO Management	Condition Monitoring	DOWNTIME REDUCTION
SUPERVISORS		PM Effectiveness Analysis	Time-Based RCA	PM Optimization	Backlog Management	MAINTENANCE COST REDUCTION
PLANNERS & SCHEDULERS		Workflow Process Analysis	Graphical RCA	MRO Kitting	Standard Work Instructions	NPT/OT COST REDUCTION
TECHNICIANS & OPERATORS		Competency Analysis	Fault Tree Mapping	5S	Daily Management	DEFECT ELIMINATION

Governance

Project Management

iBL

D.M.A.I.C.

The DMAIC method organizes the improvement process, with each phase designed to drive a successful project outcome. In the Define phase we charter the improvement project relative to business objectives. The Measure phase allows us to establish a performance baseline in order to evaluate solution effectiveness. The Analyze phase is where we identify the causes of variation or sub-optimized reliability that must be resolved in order to improve the measured performance. In the Improve phase we deploy the countermeasures and solutions, recommended through analysis, which will drive performance towards the chartered target condition. Finally, during the Control phase we will audit solution effectiveness and adjust those management systems that are required to sustain performance improvements.

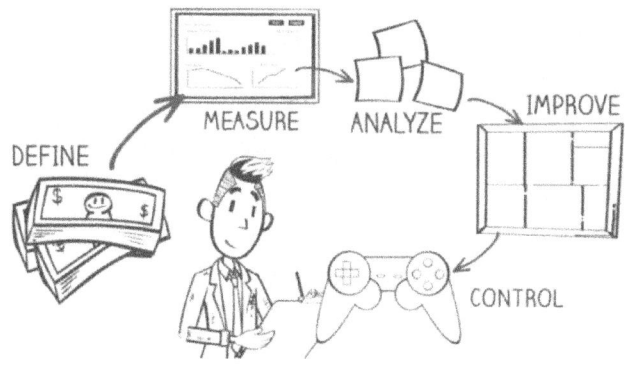

Figure 2 - DMAIC Project Management Phases

Define Phase: Selecting Project Objectives

The define phase of the DMAIC improvement process model is where we apply the lesson's learned from Volume 1 of the *Six Sigma Design for Reliability*. The goal of the define phase is to gain Top Management approval for the improvement process itself, and is commonly accomplished by way of a strategic project charter.

Strategic Project Charter

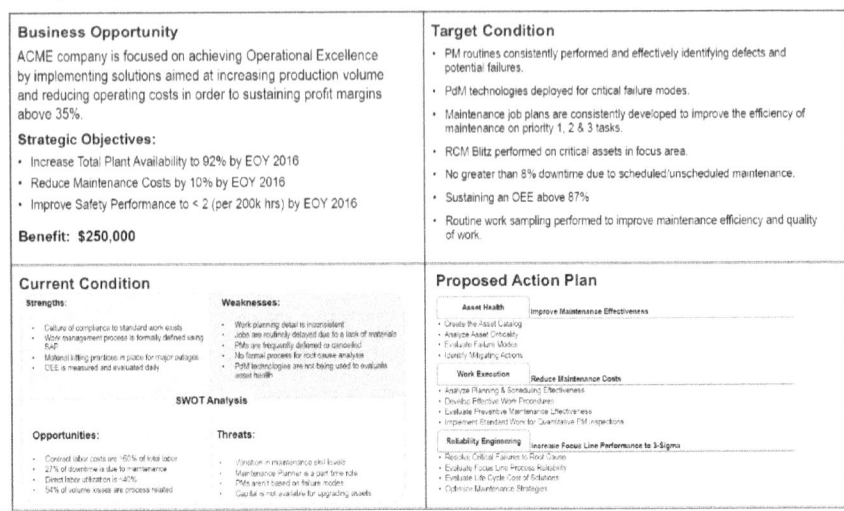

Business Opportunity	Target Condition
ACME company is focused on achieving Operational Excellence by implementing solutions aimed at increasing production volume and reducing operating costs in order to sustaining profit margins above 35%. **Strategic Objectives:** • Increase Total Plant Availability to 92% by EOY 2016 • Reduce Maintenance Costs by 10% by EOY 2016 • Improve Safety Performance to < 2 (per 200k hrs) by EOY 2016 **Benefit: $250,000**	• PM routines consistently performed and effectively identifying defects and potential failures. • PdM technologies deployed for critical failure modes. • Maintenance job plans are consistently developed to improve the efficiency of maintenance on priority 1, 2 & 3 tasks. • RCM Blitz performed on critical assets in focus area. • No greater than 8% downtime due to scheduled/unscheduled maintenance. • Sustaining an OEE above 87% • Routine work sampling performed to improve maintenance efficiency and quality of work.

Figure caption (within image): Current Condition / Proposed Action Plan (SWOT Analysis)

Figure 3 - Strategic Project Charter

The strategic project charter, often referred to as the "A3 Charter" because the entire project definition can be communicated using a single piece of A3 sized paper, defines the connection your improvement project has to strategic business objectives. The charter also defines the assumed current condition that is prohibiting the business from realizing the desired financial performance level, the target condition, which conveys the types of potential improvements that need to be made in order to achieve the desired performance level, and the proposed action plan or general outline of steps you will take in order to deploy the improvement process.

Before we continue with the construction of the strategic project charter, let's take a moment to review some of the key discussion points from Volume 1.

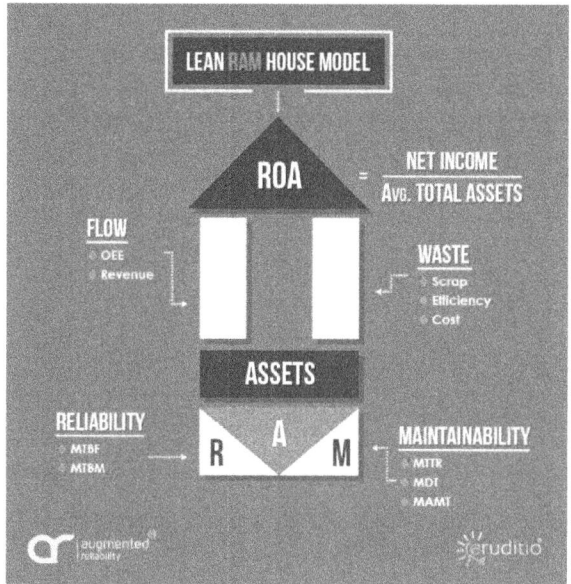

Figure 4 - Lean RAM House Model

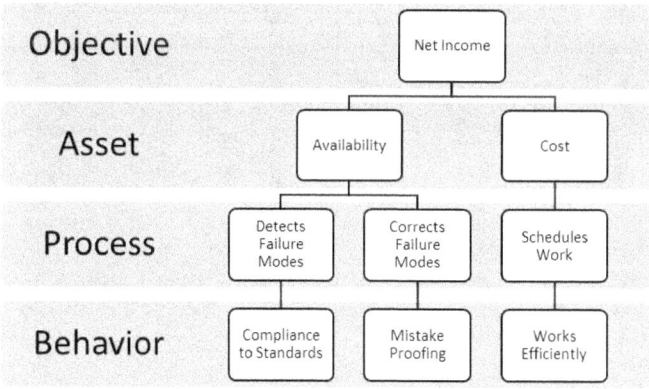

Figure 5 - Performance Metrics Structure

The Lean RAM House Model (Figure 4) helps us explain improvement priorities, with our first focus on stabilizing the asset base itself through improvements in *Reliability*, *Availability*, and *Maintainability*.

Our improvement objectives must align asset performance expectations, risk and management plans to business objectives. To facilitate this ideal, we will use a common performance metric structure

(Figure 5) to outline the linkages between asset performance and strategic business objectives, as well as, the connection between management processes and individual employee behaviors.

We must understand how <u>effectively</u> assets are performing within the desired demand period, relative to the Theoretical Capacity (Figure 6). This is particularly important because it helps us understand the difference between asset-related and process-related losses relative to the desired value generation (e.g. Theoretical Capacity).

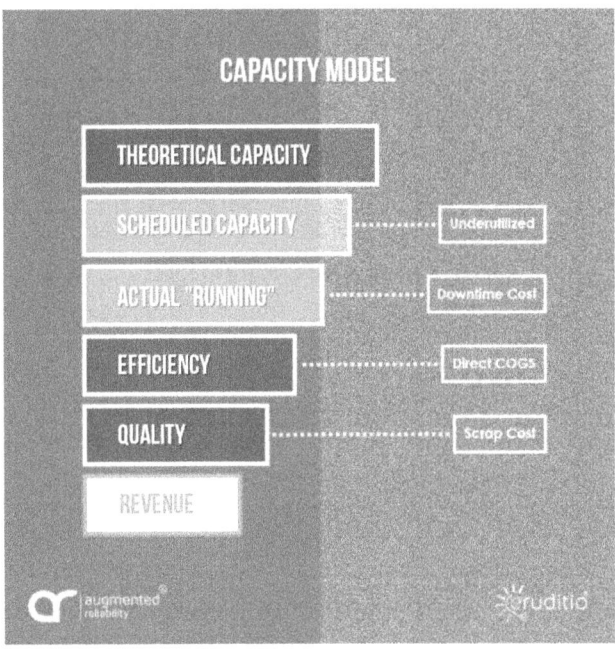

Figure 6 – Capacity Model

The realization of asset value depends greatly on the expectations of the asset stakeholders, and how these stakeholders change over the lifecycle of the asset. Each and every physical asset follows the same general lifecycle: Design, Procurement, Installation, Operation, Maintenance, and Disposal. As the asset transitions from one phase in the asset lifecycle to another so does the period of asset accountability and the definition of value and stakeholder expectations.

Strategic Project Charter

The strategic project charter consists of four key pieces of information in order to clarify the improvement opportunity, demonstrate how your project aligns with business objectives, and build confidence among stakeholders. The overall purpose of the charter is to secure Top Management's commitment for project resources, budget and execution.

Business Opportunity

The first component of the project charter is known as the "business opportunity" statement. Also known as a "mission statement" or "jumping off point", this piece of information briefly describes the strategic business objective and lists the measureable and time-bound project objectives that align to the business needs.

The more effective business opportunity statements commonly include contextual ties to multiple stakeholders. After all, your business has multiple functions that co-exist in order to achieve the business plan. When structuring your business opportunity statement, take the time to meet with various stakeholders and capture ideas or responses to these questions:

1. How does improving process efficiency create a strategic competitive advantage for our company?
2. If we improve the quality of our product, or service, what competitive advantage will it create?
3. How much additional capacity or capability can be gained by improving the reliability or availability of plant assets?
4. Will an improvement in asset or process reliability be seen as a direct improvement in workplace safety?
5. Can our improvements in asset reliability have a positive impact on environmental performance expectations?

Your business opportunity statement should be brief, unambiguous and inspire action. Using the responses provided to your questions, craft two or three statements and share them with Top Management to see which statements stakeholders find engaging. You'll know which statement is the most engaging by the number of questions you get, about the statement. If Top Management begins to ask questions like, "where do you see the biggest gap", or "do you already have a plan to meet this objective", you'll know you're on the right track towards inspiring action.

With a clear and inspiring business opportunity statement, and a list of objectives determined from your project selection efforts, the last slice of information you need to include in this first component of the project charter is the "Benefit". The business opportunity benefit refers to the quantifiable advantage to the business created as a result of your improvement project. This also implies that without your project the business will fail to realize the advantage. The benefit within your charter is commonly referred to as the "business case", however, because we are uncertain as to the amount of money or capital that will need to be invested in order to implement solutions, the benefit is only the *potential* savings, gains or contribution margin discovered in your pre-project planning.

There's no need to be conservative here. State the strategic advantage that could be realized from your project. Once you have your project underway, you can provide a more detailed business case analysis, including the payback period, to Top Management and other stakeholders.

Current Condition

The second component of the strategic project charter is often referred to as the "Current Condition". The current condition captures the perceptions, assumptions, and conclusions – right or wrong – reported by the business stakeholders during your pre-project planning efforts. The current condition is <u>not</u> intended to be a gap analysis. Instead, the current condition should reflect the common understanding of the problem needing to be solved, what strengths or benchmarked practices can be leveraged to achieve the desire outcomes, how the success of the improvement project will be measured, and the obstacles or challenges you and others may have identified that threaten the business' ability to realize the strategic advantage inherent within your project.

A SWOT Analysis is a common method used by continuous improvement professionals, and business planning experts, to address and document the pieces of information contained within the current condition portion of the charter.

Guide to S.W.O.T. Analysis

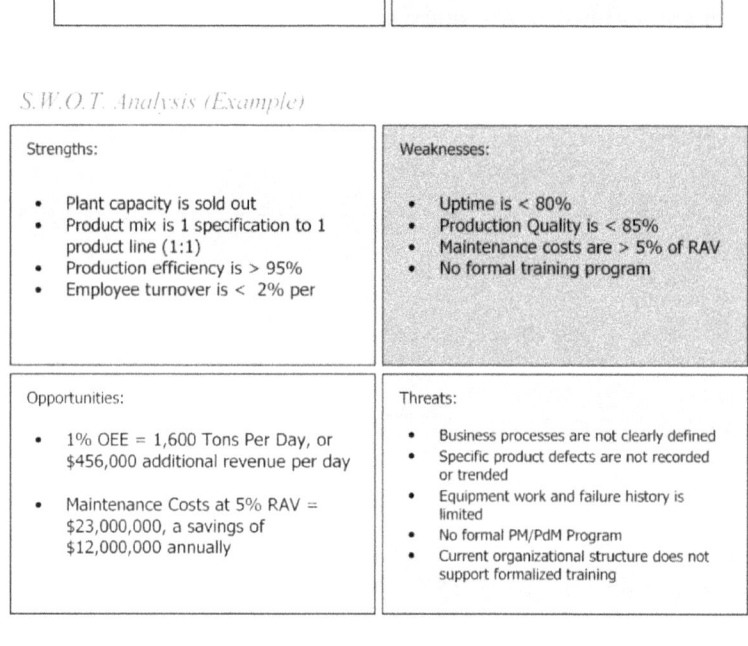

Strengths:	Weaknesses:
Consistently perform business practices that meet the desired Mission.	Current business practices that prevent the Mission from being achieved.
Consistent performance levels that reflect Mission values.	Business performance levels that are not consistent with Mission values.
Opportunities:	Threats:
Opportunities that will exist if the Global Company can overcome known weaknesses.	External or internal limitations or competitive pressures that may prevent improvement to meet desired Mission.

S.W.O.T. Analysis (Example)

Strengths:	Weaknesses:
• Plant capacity is sold out • Product mix is 1 specification to 1 product line (1:1) • Production efficiency is > 95% • Employee turnover is < 2% per	• Uptime is < 80% • Production Quality is < 85% • Maintenance costs are > 5% of RAV • No formal training program
Opportunities:	Threats:
• 1% OEE = 1,600 Tons Per Day, or $456,000 additional revenue per day • Maintenance Costs at 5% RAV = $23,000,000, a savings of $12,000,000 annually	• Business processes are not clearly defined • Specific product defects are not recorded or trended • Equipment work and failure history is limited • No formal PM/PdM Program • Current organizational structure does not support formalized training

Figure 7 - SWOT Analysis

Target Condition

A frequent mistake made by continuous improvement professionals when crafting the strategic project charter is failing to define what will change over the course of the project. The "Target Condition" piece of the charter should provide a summary of what the organization, and its stakeholders, can expect to observe, witness or otherwise recognize as problems are analyzed and solutions are implemented. Unlike the "Proposed Action Plan" which outlines the project timeline and plan of activities and resources, the target condition is meant to be used by you, the project leader, and stakeholders to demonstrate that improvement is taking place. Don't fall into the trap of simply restating the "weaknesses" in your current condition. Don't be so lazy as to merely restate the objectives from your pre-project planning. Take the time to craft a description of how things will operate differently, how people will behave differently, or how decisions will be made differently.

The target condition can also be used to state the future-vision of a process or practice. For example, if your focus line or project focus area currently does not use preventive maintenance routines (PM) to identify defects, and as a result your Maintenance organization has little to no time to plan and schedule work effectively, then you might state the target condition as:

- Failure modes will be identified for the top 20% most critical asset within the focus area.
- Preventive maintenance routines will be scheduled at an interval equal to half of the mean-time-between-failures.
- Standard work procedures will be linked to preventive maintenance work orders.
- Maintenance Technicians will be trained in the new preventive maintenance practices.
- Weekly, random audits will be conducted in order to identify continuing education requirements, and evaluate compliance to standard work.

The overall intent of the target condition is to build awareness of the changes that will likely occur as a result of your project. In doing so, Top Management and stakeholders will be able to recognize that changes are happening and proactively identify organizational change risks that threaten the success of your project. Using the previous example, your Human Resources Manager may recognize that because Maintenance Technicians are accustom to planning their own work, and that standard work procedures have not yet been introduced in the plant culture, many Technicians may resist these changes and compliance may be difficult to manage. As a result, the Human Resources Manager would likely engage with you to develop a resistance management strategy

Proposed Action Plan

The last piece of information that must be communicated within your strategic project charter is the "Proposed Action Plan". There are two camps of experts in this discussion, those who believe less is more, and those who believe that detail is the key to understanding. Here's the dividing line between the two camps. The proposed action plan _must_ convey your plan of project activities, the resources needed in order to complete each scheduled activity, and the overall project timeline. A Gantt chart – a type of bar chart, adapted by Henry Gantt in the 1910s, that illustrates a project schedule – is too large to read within the action plan portion of an A3 sized charter. On the other hand, a bulleted list of activities that does not adequately communicate resource requirements and a timeline for completion will not effectively help Top Management and stakeholders prioritize activities within your project.

At this stage of the improvement process, it is unlikely that you will be able to confidently construct a project plan and schedule without knowing what the root causes of performance losses are and what you need to do in order to mitigate them. So, within the define phase, less detail is the preferred approach. Later on in the improvement process, once you have prioritized and analyzed the root causes of performance variation, you can and should come back to the project charter and update the proposed action plan with a 30-60-90 day outlook of project activities and resource requirements. In doing so you will continue to utilize the project charter as a communication tool over the duration of your project to keep Top Management and stakeholders informed of progress being made, potential threats to the project, as well as, the upcoming activities.

Measure Phase: What Process Variables are Impacting Performance?

With your improvement project chartered by Top Management, the next step in the DMAIC methodology is to establish a performance baseline and answer the question, "What process variables are impacting performance?" We are establishing a baseline for two reasons. First, we want to clarify which variables are currently impacting performance, and to what magnitude, so we can prioritize our analysis and improvement efforts. It's the "biggest bang for the buck" idea we are after first. Second, the baseline will serve as a comparative value when evaluating the effectiveness of your solutions. In complex projects, like asset reliability improvement projects, where multiple solutions are required in order to achieve the desired target condition outlined in the project charter, we must have a way to evaluate each individual solution and verify that each solution is leading us towards the end goal.

In the Measure phase we will use one of four techniques to establish the performance baseline:

1. Value Stream Mapping – This technique is used to document the flow of work from the time a customer order is received, through the primary process, and ending at the supply of materials used within the process. The term "value stream" refers to the tasks or activities performed within the process that engineer, manufacture, buy, store, install, operate or maintain an item in order to generate the desired Net Income for the business. The Value Stream Map is a visual representation of tasks or activities that make up the product or service produced by the process, and how each activity impacts cost, customer service, or other measures of value within the business.

2. Process Workflow Mapping – A technique used similarly to Value Stream Mapping but at a lower level within the overall

value stream. The Process Workflow Map illustrates the steps needing to be performed for one, singular activity within the value stream. A Process Workflow Map often depicts the flow of labor, materials or information through a process in order to better understand how these variables impact the process, and subsequently the value stream.

3. Reliability Block Diagram – The Reliability Block Diagram is also a subset of the Value Stream Map. This technique adds asset performance to the list of possible variables. In a Reliability Block Diagram we can see how assets interrelate relative to their designed or desired functions – manner in which an asset adds value. Using statistics to quantify the probability of Reliability, Availability or Maintainability, the Reliability Block Diagram illustrates the impact each asset has on the overall likelihood that the process, or process step, will meet the desired performance expectation.

4. Graphical Analysis – The technique most commonly used to measure variability within a process, process activity, or process step when the sources of variation are already known is called Graphical Analysis. Pie Charts, Pareto Charts, Histograms, Control Charts, and Linear Regression Charts are all examples of this technique. These charts provide a graphical representation of collected data in order to quantify the impact of one variable over another, or illustrate the relationship between a single variable, the 'X', and process performance, the 'Y'.

The Six Sigma Design for Reliability

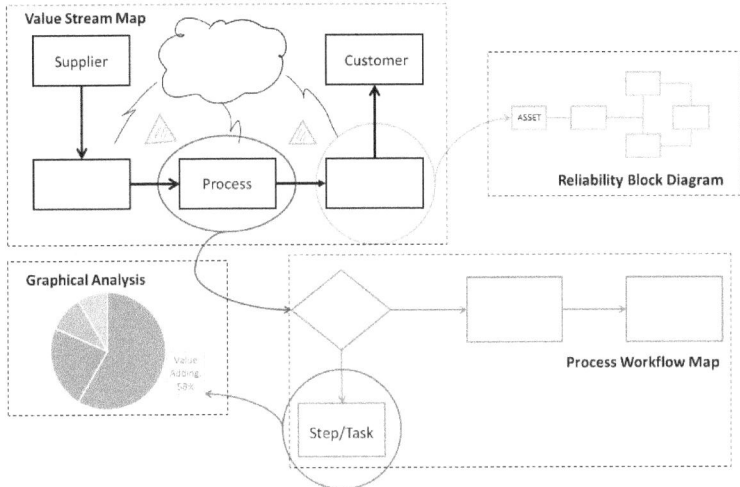

Figure 8 – Cascading Measurement Techniques

PROCESS MEASURES

When using a Value Stream Map to quantify the sources of process variation **Time** is the most common measurement, and "Takt Time" is the initial prioritization criteria.

$$Takt\ Time = \frac{Available\ Production\ Time}{Required\ Quantity\ of\ Output}$$

Takt Time is a measurement of your process' ability to produce, operate or function relative to the demanded output. The Takt Time formula illustrates how demand is calculated in the denominator, with the desired lead-time or turnaround-time as the numerator. For example, if your Customer has a daily order quantity of 1,000 Units and the available production time within your process is 12 Hours (720 minutes). This equates to a Takt Time of 0.72 minutes per unit produced – 1 unit must be produced every 0.72 minutes in order to meet demand.

$$Takt\ Time = \frac{720}{1000}$$

$$Takt\ Time = 0.72\ Minutes\ per\ Unit$$

Overproduction occurs when your value stream, or any individual process within the value stream, is operating ahead of Takt Time. Overproduction is a form of waste because the cost incurred for each over-produced unit will not translate into revenue, resulting in a lower Net Income.

Underproduction, similarly, is also a waste. A process or value stream that is incapable of meeting Customer demand is considered to be "sub-optimized" – not reaching its full revenue potential – and will deliver a lower Net Income because of the increased Fixed Costs with fewer units sold.

In addition to Takt Time, we can also measure extraordinary costs within the value stream. Six Sigma practitioners often refer to these as "Waste", as we saw from the Lean Ram House Model in Figure 4. A good benchmark to use when measuring the cost of waste within your value stream is 2% of the total Cost of Goods Sold.

Inventory, either raw material or direct materials, that are sitting in work in process (WIP) staging areas are an indicator of cost accrual that may be unnecessary, it may be adding to the cost of "waste". The staged WIP inventory is often a symptom of an imbalance between processes. For example, if the downstream process has a faster cycle time – the time it takes the process to transform 1 unit – than the upstream process, a WIP staging area is used to buffer the overall flow of materials and products through the value stream. These buffers enable the value stream to maintain a steady flow, compensating for those processes that are not aligned with the Takt Time, in order to meet Customer demand. WIP buffers can also be a sign of poor asset reliability.

"Inventory Efficiency" is a common metric used to draw the line between value-adding and non-value adding material costs. Inventory Efficiency is measured as:

$$Inventory\ Efficiency = \frac{Total\ Sales}{Average\ COGS\ Inventory}$$

$$Inventory\ Efficiency = \frac{1,000,000}{25,000}$$

Inventory Efficiency $= 40\ Turns$

As an example, assume your company has $1 million in sales, and the average monthly Cost of Goods Sold (COGS) inventory is $25,000. Using the equation, the company has inventory turnover of $1 million divided by $25,000, or 40 turns. To translate this into inventory days on hand we need to know the total COGS. We'll assume $250,000. Divide the total COGS by the average COGS inventory. Then, divide 365 days by that number. The answer is 36.5 days. This means under the first approach, inventory turns 40 times a year, and is on hand approximately 36 days. Is that good? It depends on whether or not you NEED the inventory on hand. If your process was more reliable would you need the inventory? If you were producing within the desired Takt Time would you still need the inventory?

Maintainability

Within the framework of the *Six Sigma Design for Reliability*, Process Workflow Mapping is most often used to measure "Maintainability". Maintainability is formally defined as the ease to which the Maintenance organization restores asset function, and subsequently asset performance. As a process, Maintainability refers to the probability that work will be performed within a specific time period, or "Standard Time". In an automobile manufacturing plant, Maintenance may be required to complete work within a 20 minute downtime window. In a power generation plant, Maintainability might be measured based on the duration of the scheduled outage. Using this definition, Maintainability is calculated as:

$$M(t) = 1 - (e^{-\left(\frac{Standard\ Time}{MTTR}\right)})$$

Let's use a simple example to illustrate this measurement method. If your Maintenance organization completes a work order in 12 hours on average, then your Mean-Time-To-Repair (MTTR) is 12 hours. Now, if we assume that Production wants to limit downtime to less than 4 hours in order to minimize WIP inventory levels and sustain Takt Time, our Maintainability equation would look something like this:

$$M(t) = 1 - (e^{-\left(\frac{4}{12}\right)})$$

$$M(t) = 0.28346$$

$$M(t) = 28.35\% \ Probability\ of\ being\ maintained\ on\ time$$

Day In the Life Observation (DILO) Worksheet			
Time Lapse (minutes)	Observed Activity	Value Added	Non-Value Added
0	Cleaning workbench	x	
30	Obtain work order for job no. 2345-04	x	
45	Searching for materials		x
5	Travel to job site	x	
0	Begin work order no. 2345-04	x	
5	Return to shop to obtain tools		x
30	Restart work order no. 2345-04		x
20	Completed work order no. 2345-04	x	
	Total Activities Performed	8	
	Total Value Added Activities	5	
	Total Non-Value Added Activities	3	
	Percentage of Value Added Activities	63%	
	Total Time Elapsed (minutes)	135	
	Duration of Value Added Activities	55	
	Duration of Non-Value Added Activities	80	
	Percentage Net Available Labor Utilization	41%	

Figure 9 - DILO Worksheet

As a leading indicator of Maintainability, we can also measure direct labor utilization, most commonly referred to as "Wrench Time". Wrench Time is the duration of time, as a percentage of the total available labor hours, in which a Technician is performing one of the following tasks:

- Inspecting an asset,
- Cleaning an asset,
- Testing asset parameters, such as vibration, temperature or pressure, and
- Correcting an abnormal asset condition (e.g. a "defect").

A good benchmark for Wrench Time is 55%-65% of the "Net Available Time", which is straight-time or payroll hours minus time allocated to training, meetings, meals, breaks and other indirect commitments.

Other variables that impact Maintainability include:

- Stockouts – maintenance materials and spare parts that are not available when requested,
- Walking – time consumed by traveling to and from the work site and the workshop, storeroom, tool crib, etc.,

- Waiting – time consumed while the technician waits for the asset to be released by Operations and prepared for servicing, and
- Planning Accuracy – the variance between the estimated duration of work and the actual duration of work.

Reliability

Turning our attention towards the Reliability Block Diagram, as a measure of asset performance within the value stream, we can also calculate variation. Similar to how we calculated Maintainability, Reliability is a measure of probability or likelihood that the asset will perform as desired, within a given period of time. In the context of Reliability, the time comparator is referred to as "Mission Time" – the demand period or desired scheduled production time. The equation for calculating Reliability is as follows:

$$R(t) = e^{-\left(\frac{Mission\ TIme}{MTBF}\right)}$$

Again, we'll use a simple example to explain this measurement method. If your value stream needs to operate for 30 days, or 720 hours, without interruption in order to meet Customer demand, but, you experience a process fault or asset-related failure on average every 125 days, or 3000 hours (e.g. the "Mean-Time-Between-Failure"), then the Reliability of your process would look like this:

$$R(t) = e^{-\left(\frac{720}{3000}\right)}$$

$$R(t) = 0.7866$$

$$R(t) = 78.66\%\ Probability\ of\ operating\ without\ interruption$$

INCORPORATING THE "VARIABLES"

Finally, we come to Graphical Analysis and the types of measures we can use to illustrate the priorities for our next phase of the improvement process. The Analyze Phase.

Graphical Analysis allows us to visually compare the measures we collected from the value stream, process workflow or reliability block diagram and compare them to variables in order to determine which variable, or small set of variables, has the greatest impact on one of these outcomes.

In Graphical Analysis we call the variables the "X's" because they generally appear on the X-Axis of a Pareto Chart. We call the Takt Time, Maintainability, and Reliability outcomes the "Y's", again because this is the axis upon which they appear in our charts. The "Y's" could also be project objectives, like cost, quality or customer satisfaction.

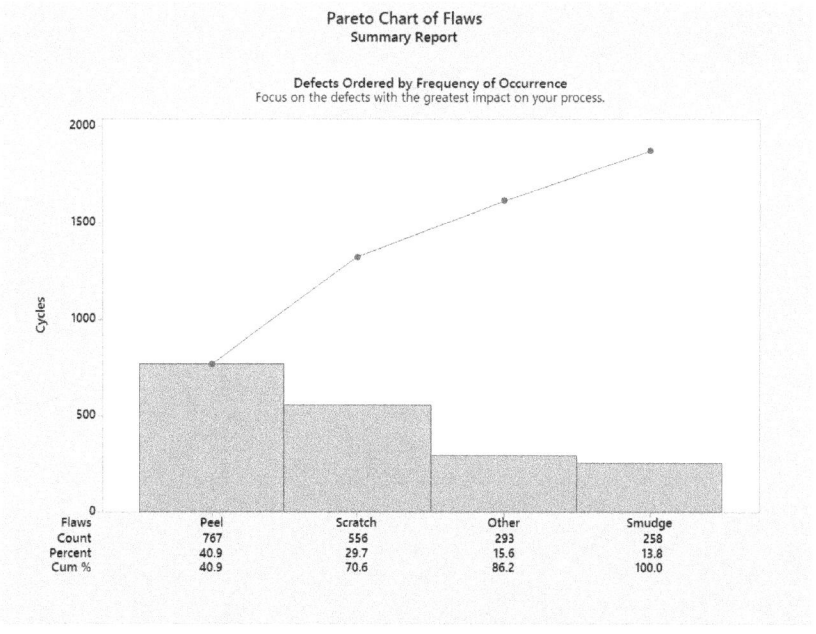

Figure 10 - Pareto Chart

Pareto Charts are the most commonly used form of graphical analysis, and where our "Xs" and "Ys" reference originated. Pareto Charts illustrate the *frequency* at which a variable occurs. Pareto Charts typically begin at a high level with the data collected from your value stream map, process workflow map, or reliability block diagram. However, in order to adequately prioritize these variables and execute the Analysis phase of the DMAIC methodology, we must drill down to the lowest common denominator.

For example, using Figure 10 as a starting point, let's assume that Category 1 happens more often than other process variables. And we'll assume that Category 1 is "Peel" – the paint peeling after application. Is there more than one manner of "Peel"? If there is more than one manner of failure, then we need to collect data on each type of failure (e.g. Failure Modes) and Pareto these variables to identify the leading cause of "Peel". It is common to have three or four layers of Pareto Charts before you arrive at the most common cause of less than ideal process performance.

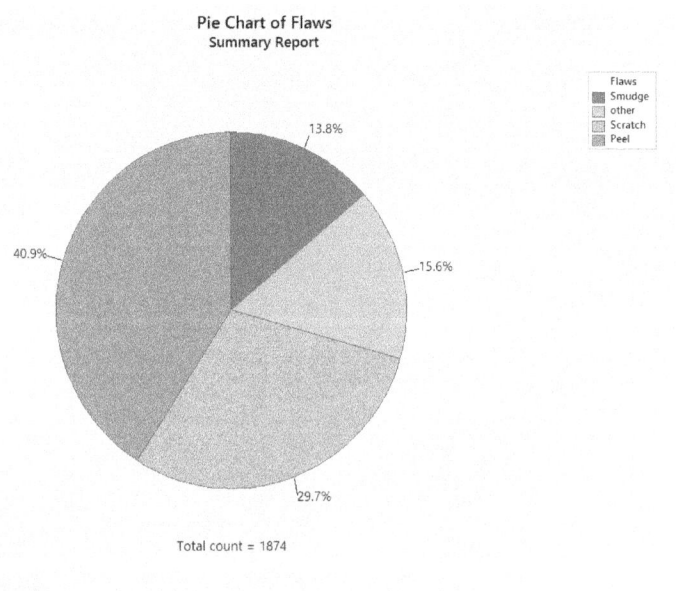

Figure 11 - Pie Chart

Another commonly used form of Graphical Analysis is the Pie Chart. Pie Charts also help us prioritize variables, but in a different way than Pareto Charts. Pie Charts are used to distribute proportional data, such as value-adding vs. non-value adding time. Time, in this case, is the cause of less than ideal process performance and we would use the Pie Chart to determine the tasks or activities that are consuming the largest portion of time.

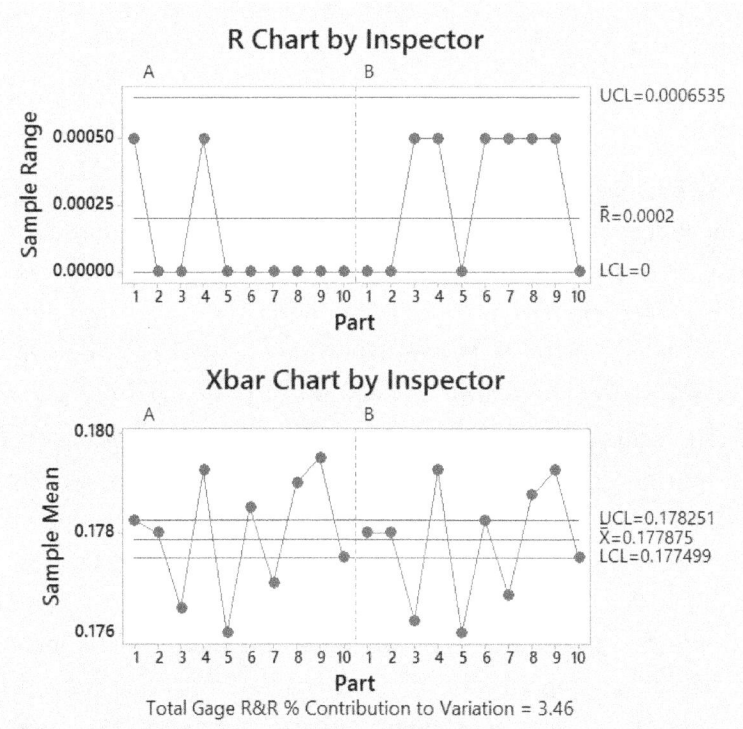

Total Gage R&R % Contribution to Variation = 3.46

At a higher level, Control Charts can be used to quantify the impact of each variable relative to time, but not as an aggregate proportion. Control Charts are used by Six Sigma practitioners to identify the *point in time* when a process is above or below the desired level of performance in order to determine the time of process variability, or the frequency of variability over time. Control Charts, when used with other forms of graphical analysis, can help us focus on the narrow few instead of the trivial many. Those few events in time that are above or below the upper and lower control limits.

Figure 12 - Control Chart

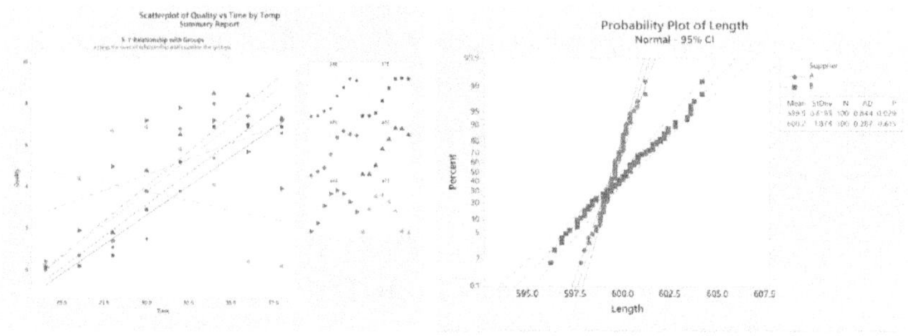

Figure 13 - Linear Regression Charts

Finally, we come to Linear Regression Charts, sometimes referred to as "Scatter Plots" or "Probability Plots" because of the scattering of data nodes within the chart, and your ability to predetermine an outcome based on the trend. Linear Regression Charts are most commonly used with the Reliability Block Diagram in order to illustrate a correlation between failure events within a snapshot of time. Each dot within the chart represents a failure in time. When we connect these events using a Best Fit Line – a straight line that passes through some, or all of the dots and intersects the Y-axis – we can then calculate the slope of the line as the rate of regressions, or the probability of recurrence. The slope of the line, also known as *Beta*, tells the story. If the slope of the line is greater than zero than we have an indication of increasing regression. If the slope trends in the opposite direction, downward, the slope is less than zero, indicating a decreasing rate of regression over time. And, if the line is flat, meaning only one dot fits to the line, then we can say that the events are not related and are randomly distributed.

As the last step in the Measure Phase, graphical analysis helps us determine how our data relates to process performance. When evaluated, the graphed data enables us to prioritize how we will tackle the analysis of root causes in the Analyze Phase.

Analyze Phase: What is Causing the Variation?

The Analyze phase is where we identify the causes of variation or sub-optimized reliability, maintainability and availability that must be resolved in order to improve the measured performance. This is typically where we begin to deploy techniques associated with "Root Cause Analysis" – a systematic approach to problem solving. Our goal in the Analyze Phase of the improvement project is to define a clear roadmap for improvement. A roadmap that addresses the highest priority variables first.

TIME-BASED ANALYSIS METHODS

"Time" methods are preferred when analyzing variables relative to a time sequence. These methods help you determine if causal chains are in fact interrelated in time. Time-based methods help organize seemingly random factors into a logic sequence or scenario to explain how variation occurs.

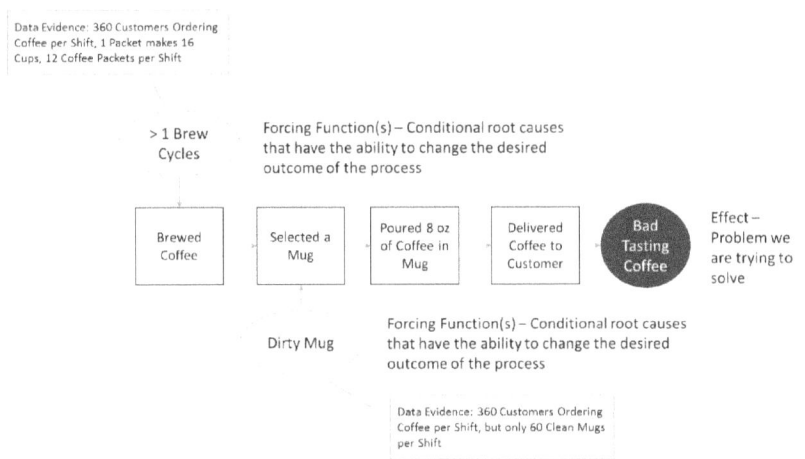

Figure 14 - Sequence of Events Method

The best method to use when trying to identify the importance of each contributing factor in the causal chain is Sequence of Events. This method displays a horizontal causal chain, relative to time, leading up to the specific problem needing to be solved. It is common, as well, to document the events in time after the problem as these factors may have led to the frequency at which the problem occurs.

Events should be written in a way that states what happened, not a condition, conclusion, or suspected circumstance. Then, add the evidence collected to the diagram to validate the primary event sequence.

Within the Sequence of Events method, we also need to identify the conditional circumstances, such as asset parameters or environmental changes that could have contributed to an event or led to the event causing the variation in process performance. Some practitioners will also refer to these circumstances as "Forcing Functions".

When dealing with time-related problems in which various contributing conditions or event timelines exist, it is best to expand the Sequence of Events by using the Event and Causal Factors Analysis method. This method helps you determine the relationship in time of primary, secondary, and conditional causes, especially if you are expected to process a large volume of data, evidence, or eyewitness accounts that appear to be unrelated to the physical events that led to the undesired condition.

TREE-BASED ANALYSIS METHODS

"Tree" methods are used to examine the undesired effects or sources of variation within a time sequence or physical system, such as the introduction of quality defects and equipment breakdowns. Tree methods present the possible causes identified by your data collection in branching scenarios that represent the logical ordering of known factors, with each scenario then evaluated using evidence to determine solution selection.

The Five Why method (a.k.a. "5-Whys") is a basic problem solving tool that evaluates a single set of causes by asking why each event or factor occurred in a chained progression, typically from top to bottom. The reason for the "5" in the "Five Why" is to ensure that human and potentially systemic root causes are documented in the causal chain. Stopping before the 5th "Why" may only capture the physical events that occurred and may not provide enough detail for effective solution selection.

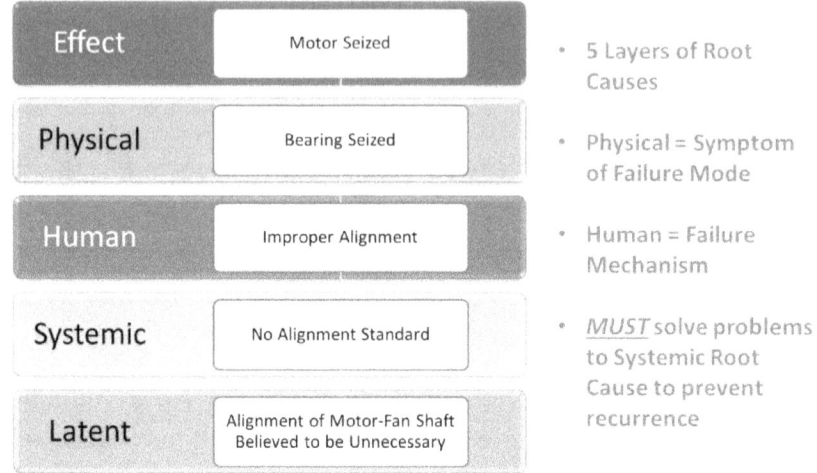

Figure 15 - 5 Layers of Root Causes (e.g. "5-Whys")

Another tree-based method is the Fault Tree Analysis (FTA). Fault Tree Analysis is simply a branched Five Why. When you are faced with a multi-faceted problem that could have long causal chains, the FTA method is the preferred approach in order to achieve a common understanding of all of the major factors that could have contributed to the system's undesired outcome. This is an advanced method and is a better tool to use than the Five Why method when trying to solve complex, equipment-related problems. We must remember that when dealing with equipment-related problems we always have a minimum of two (2) causes that exist at the same point in time, a conditional cause and an actionable cause. This means that directly under your effect or problem needing to be solved, you will have at least two (2) causal chains. For this reason alone, the Five Why method is inadequate.

The Logic Tree Analysis (LTA) method is used to examine the various scenarios represented in a fault tree using logic to determine if causal chains are independent or interrelated.

This method uses "And" statements to illustrate that two (2) or more chains are related in time and both must occur to cause the problem. We love to see "And" statements because it reduces the number of solutions that have to be implemented. When you have two (2) causal factors that are linked by "And", you only have to eliminate one (1) to effectively prevent the source of variation from occurring again in the future.

"Or" statements are used to illustrate the opposite, that each chain or branch independently causes the problem with no relationship to other factors. With an "Or" statement, you must implement a solution for each cause in order to prevent reoccurrence.

TRANSPARENCY-BASED ANALYSIS METHODS

"Transparency" methods are used to proactively identify product design, safety, quality, or reliability problems that have the potential to impact process outcomes. These methods create visibility of unknown relationships between humans, machines, and materials, as well as, control mechanisms, like standard operating procedures and preventive maintenance routines, that may be ineffective in mitigating risk.

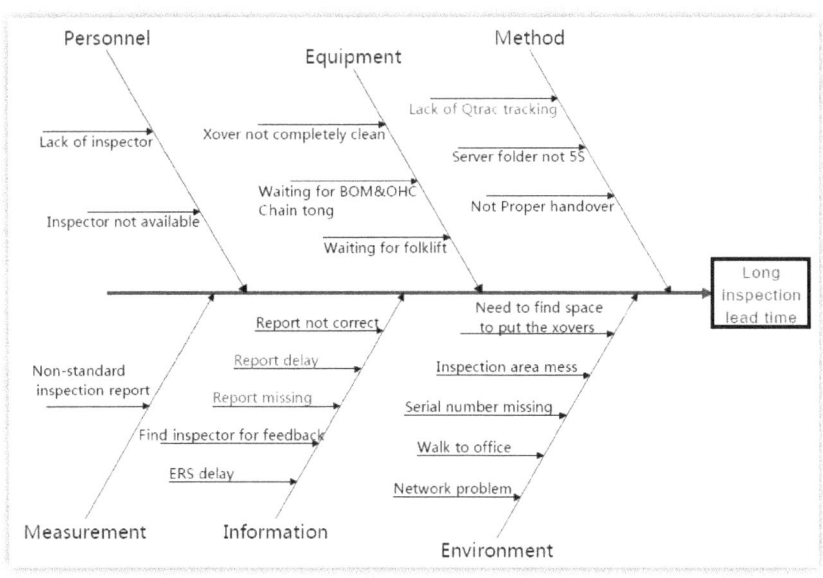

Figure 16 - Fishbone Diagram

A Cause and Effect Diagram (a.k.a. "Fishbone Diagram") is a basic brainstorming tool used to illustrate the relationships of various causal factors that may contribute to the problem, or "effect". Most practitioners facilitate this brainstorming process by creating four (4) branches, one (1) for each causal factor category. We call these branches the "4-Ms", which stand for Machine, Methods, Materials, and Man. This helps you to organize the data collected during the Measure Phase to better understand what causal factors need to be analyzed further using the Failure Mode, Effects, and Criticality Analysis (FMECA) advanced transparency method.

The simplified FMECA is used to identify likely failure modes in a top-down approach from system to component. We call it "simplified" because this form of Failure Mode Analysis (FMA) stops at the component level. Instead of examining the individual failure modes and effects of replacement spares such as fasteners, gaskets, and springs, the simplified method looks at the relationship of these parts to their parent component or machine as the potential causes of failure. The relationship between component, part, and problem is what we call the "Failure Mode", and the relationship between problem and cause is known as the "Failure Mechanism". From here, we can identify if a new risk mitigating action, or control, is needed to prevent the failure mechanism from occurring.

Process Function	Potential Failure Mode	Potential Failure Effects	S E V	Potential Causes	O C C	Current Process Controls	D E T	R P N
Pump fresh waster from the reservoir to the cooling tower basin at a rate of 1800 gpm when the basin low limit switch is open.	The motor drive end bearing seized	Motor amperage spiked before pump stopped		due to over lubrication		12 grams of grease is the standard but we don't have a good way of ensuring only 12 grams is added, or is needed each month		
	The basin low limit switch failed to open	Pump didn't start when basin level fell below lower limit		due to corrosion		Postmortem - Limit switch cowl was cracked as found during failure investigation		
	The pump impeller was damaged	Low pump discharge pressure was noted during last inspection		due to improper installation		No standard maintenance procedure for installing pump impeller or testing pump operation afteroverhaul		
	The pump coupling key backed out	We didn't know we had a problem until we opened the pump		because it was the wrong size		Postmortem - Coupling key was found dislodged during failure investigation		

Figure 17 - Failure Mode Effects and Criticality Analysis (FMECA)

One of the advantages of starting your analysis with the Fishbone Diagram, as a cause and effect method, is that it helps the team gain a common understanding of the big picture issues, especially if team members came to the problem solving event prepared to contribute ideas based on their cross-functional perspectives.

The Failure Mode, Effects, and Criticality Analysis (FMECA) method helps quantify the risk priority of each identified failure mode within the cause and effect relationship. A FMECA analyzes risk relative to how severely the failure mode impacts process outcomes, such as Takt Time, the probability that the failure mode will occur again in the future (e.g. Reliability), and how likely it is that your organization will detect the onset of the failure mode and correct it before the effect is realized by the organization (e.g. Maintainability). The sum of these three (3) risk factors is known as the Risk Priority Number (RPN) of the failure mode and can be used to prioritize solution selection. This is particularly valuable when comparing the effectiveness of current controls and potential solutions.

Improve Phase: Deploy Your Solutions

In the Improve phase we deploy the countermeasures and solutions, recommended through analysis, which will drive performance towards the chartered target condition.

PROCESS PERFORMANCE IMPROVEMENT TECHNIQUES

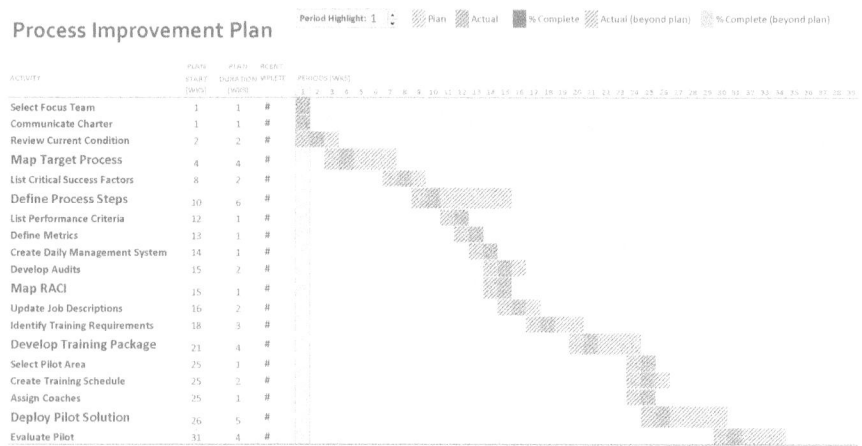

Figure 18 - Process Improvement Plan

Looking back at Figure 8 and our cascading measurement techniques, before we begin our discussion on process improvement we must remember that our efforts in the previous two phases of the DMAIC methodology have helped us narrow our focus to *the* process or sequence of steps that have the greatest impact on overall value stream performance. The following discussion is then intended to deploy solutions that will eliminate the root causes of process losses, not to re-engineer the value stream in its entirety.

In this chapter we will focus on three critical techniques within the Process Improvement Plan:

1. Mapping the "Target" process,
2. Defining role responsibilities using a RACI or RASI chart, and
3. Deploying a "Pilot" solution.

Process improvement, using the listed techniques, can be achieved in 26 weeks within an organization that are ready for change. However, if paths of resistance to change are not effectively managed, your improvements could take significantly longer. We will discuss how to evaluate resistance to change in a moment, and how to manage resistance to change in a subsequent volume of the *Six Sigma Design for Reliability* series.

Mapping the Target Process

With the current condition of your process already defined, and root causes of less than ideal performance documented, our next step is to define how the process *should* function relative to a known standard or benchmark. This is the first iteration of the new process workflow.

The "Should-Be" process workflow is often an existing standard found within your organization, with the current or "As-Is" process merely a derivative of the known standard. When mapping the process our goal is to compare how work is done today relative to how work should be done if your organization was complying with the established standard. In many cases, the current process is the result of:

- Insufficient information where the work is being performed,
- People lacking the necessary skills needed to perform the work,
- Decisions being made outside of the work being performed, and
- Misalignment between the work and the current infrastructure supporting the work.

When mapping the "Should-Be" process, focus on seven key principles to streamline the workflow process and thereby achieve significant levels of improvement in work quality, time management or cost:

1. The workflow process should be focused on outcomes, not tasks,
2. Integrate information processing into the work that produces the information,
3. Distribute work by roles, not individuals,

4. Make decisions where the work is performed,
5. Build controls into the process,
6. Capture information once and at the source, and
7. Link parallel activities within the workflow instead of just integrating their results.

The workflow process that represents your solution to remove or obliterate the non-value adding work is known as the "Target" or "To-Be" condition. In this iteration of the workflow process the current condition is modified in order to create a step-change towards the "Should-Be" process workflow. Because the inefficient non-value adding work was created as a result of the "Should-Be" process lacking the proper infrastructure, decision making framework, information management and skills, it may be necessary to implement an interim process that aligns with your chartered project objective but falls short of the ideal standard of practice until such improvements can be made.

Defining Role Responsibilities

Now that you have an implementable process workflow that incorporates solutions identified during the Analyze Phase, and aligns process outcomes with the chartered project objectives, it is time to begin mapping organizational roles to process workflow responsibilities. This critical step in the improvement process is meant to remove ambiguity, which may be causing the current waste-level, and drive new standards of practice through organizational accountability and behaviors.

RACI Charts

Steps will be highlighted if A and R are not assigned.

R:	Does the step
A:	Accountable for the step
C:	Consulted for decisions
I:	Informed when the step is completed

Project: Process Improvement

PROCEDURE Work Control

DEPARTMENT Maintenance

STEP	DESCRIPTION	Supervisor	Planner	Technician	Stores Clerk	Operator	Engineer
1	Request Work	A	I			R	C
2	Approve Work	R	A			I	C
3	Prioritize Work	A	R				C
4	Plan Work	I	R	C	I		A
5	Schedule Work	A	R		I		C
6	Assign Work	R	A	I			C
7	Complete Work	A		R		I	C
8	Close Work	A	R		I		I

Figure 19 - RACI Chart

A RACI or RASI Chart is the ideal tool to use when defining process role responsibilities. The RACI Chart helps us see where roles may be overloaded with responsibilities, or where information and data need to be disseminated in order to efficiently make decisions within the workflow. RACI, as an acronym, stands for: Responsible, Accountable, Consulted, and Informed. For each step or task within the process workflow, including decisions, we must define the R.A.C.I. roles.

Responsible roles for each step may be many, but should align to the current job descriptions and fundamental duties of each role. For example, when it comes to assigning work within your process it is recommended that you assign the natural supervisory roles to this type of process step. Responsible means that the assigned roles are the persons expected to execute the step as you have defined it.

Accountable roles within the RACI Chart are unique in the fact that there can be only one per step or task. Assign accountability to the managerial role that has financial or measureable authority for the process outcomes.

Consulted roles are those persons within your organization that have information that is not obvious or evident within the process workflow you have defined. For instance, if your process requires Maintenance work to be coordinated and scheduled, but you need Production to help identify windows of equipment availability for maintenance, then you would assign the 'C' to a Production role instead of a Maintenance role.

Informed persons within your process workflow are typically those roles that exchange accountability as the process evolves. Assigning the 'I' within the RACI Chart ensures that information technology administrators distribute the reports, dashboards, and other forms of electronic information and data necessary to drive process performance. A common example of the information exchange within a Maintenance process is when the Storeroom Clerk completes the assembly of work order material kits and notifies the Maintenance Planner/Scheduler via the work order system, and an electronic status code, that signifies that the work order is ready to schedule.

Deploying the "Pilot" Solution

Finally, we are ready to test our solution and measure performance improvement relative to the baseline we documented in the Measure Phase of our DMAIC methodology. This test is known as the "Pilot" implementation period.

The Pilot Implementation Period is typically the first 9 weeks of implementation during which the organization is actively monitoring the solution and auditing both compliance to the new process and associated practices, and process outcomes. It is important to audit both parameters. If compliance is poor, and those impacted by the changes revert back to doing things the old way, then outcomes make take longer to change. Coaches should be identified and strategically placed within the process workflow to monitor behaviors and practices, and provide immediate feedback to correct performance.

The Six Sigma Design for Reliability

Implementation Area SWOT Analysis - Pilot Area Selection

Change Ready
Change Resistant

Area & Change Risk Index	Smelting	59%	Extruding	41%	Forming	18%	Packaging	35%
Strengths								
>80% of Work is Planned	No	0	No	0	Yes	1	Yes	1
Backlog is <8 Weeks	No	0	No	0	No	0	No	0
PM Compliance is >90%	Yes	1	Yes	0	Yes	1	No	0
Weaknesses								
Maintenance workflow is informal.	No	0	Yes	1	No	0	No	0
Job plans lack sufficient detail.	Yes	1	Yes	1	Yes	1	Yes	1
MRO spares are not Kitted.	Yes	1	Yes	1	Yes	1	Yes	1
Opportunities								
Direct Labor Utilization <40%	Yes	1	Yes	1	Yes	1	Yes	1
Maintenance Downtime is >10%	No	0	No	0	Yes	1	Yes	1
Operational Efficiency is <90%	No	0	No	0	Yes	1	Yes	1
MRO Stockouts are >2 per Week	Yes	1	Yes	1	Yes	1	Yes	1
Threats								
>20 People to train.	Yes	0	Yes	0	No	1	No	1
Conflicting improvement priorities.	Yes	0	No	1	No	1	No	1
>20% of Assets are "critical".	Yes	0	Yes	0	No	1	Yes	0
No front-line sponsorship.	No	1	No	1	No	1	No	1
Technicians are satisfied with current state.	No	1	No	1	No	1	No	2
Past improvements are viewed as "failures".	Yes	0	Yes	0	Yes	0	Yes	0
Metric reinforce current state behaviors.	Yes	0	No	1	No	1	Yes	0

Figure 20 - Pilot Area Selection SWOT Analysis

To prepare for pilot implementation, it is recommended that you evaluate the organization's readiness for implementation by operating area, department, or geographic plant location. This is accomplished in five steps:

1. Identify the criteria that will be used to evaluate the current level of workflow process stability within each potential implementation area. These are the STRENGTHS upon which your process improvement solutions will gain momentum. Instability within the current process refers to a workflow process that is overwhelmed by the volume of work and has a demonstrated history of being ineffective. It is best to use metrics that measure the *consistency* of process execution when defining the area STRENGTHS. An inefficient process that is consistently followed is easier to change than a process that lacks discipline and compliance to current standards of practice.

2. Define the platform for change that has been identified via your root cause analysis efforts. These WEAKNESSES should be viewed by those employees that will be impacted by your process improvement solutions as a common need for change. If employees in the implementation area do not agree with the need for change, then resistance to change is highly likely. Resistance delays implementation and the projected return on investment.

3. List the business metrics, or key milestones associated with the improvement process that will prove that your solutions are positively impacting the strategic business objectives of the organization. Improvement OPPORTUNITIES for the business that are directly related to your project will ensure management's commitment to see solutions implemented. OPPORTUNITIES should be the *leading* indicators that create confidence in the solution and display a pathway to the ultimate goal.

4. THREATS to the success of process improvement implementation relate more often to the organizational dynamics within each implementation area. A greater number of employees impacted by the change, divided sponsors competing for improvement resources, or the perception of unsuccessful past improvement initiatives may create additional resistance to change. Evaluate how the organizational dynamics vary from area to area in order to find the most change ready implementation area. This minimizes the level of change management required by plant leadership.

5. Perform the SWOT analysis for each potential pilot implementation area by involving plant leadership in the analysis. Apply a 'Yes' response if the statement or metric is true for the area, and a 'No' response if the statement or metric is false for the area. Based on the applied responses, select the pilot implementation area using the overall Change Risk Index – a numerical representation of the level of implementation risk or resistance to change. A low Change Risk Index means that the area is more change ready. A higher index requires a greater level of leadership participation, communication planning, active coaching, and upfront training in order to successfully pilot your solutions.

Control Phase: Sustain the Gains

Change management is a structured process for leading the people side of improvement. Effective change management is when the structured process results in transforming the organization and its individuals in order to achieve the desired business objectives. There are three phases associated with this transformation:

- Preparing for change, in which stakeholders are aligned in vision and strategy based on the immediate business needs.
- Managing the change, by integrating change management activities within the project plan, actively identifying and resolving pockets of resistance through leadership, and demonstrating a commitment to winning.
- Reinforcing the change through performance accountability and leadership competencies.

Within the *Six Sigma Design for Reliability* improvement process, the purpose of the Control phase is to quickly identify the sources of non-conformance to new standards of practice and deploy the appropriate risk mitigating actions. We will dive into more detail about managing organizational change in a future volume, however, before embarking on your improvement process let's understand how to create a plan that reinforces change.

Creating the Ability to Change

Choices are something that each person has to themselves, and no other person, despite our many attempts as parents and as business leaders, can make their choices for them. People in your business make choices every day that impact the profitability, productivity, and ultimate performance of your company. Although we cannot make these choices for them, we can influence them a great deal. Communication cannot be your only strategy for leading change as it will not adequately lead people to make the decision to choose change.

One of factor that limits the results of improvement initiatives is the rate or speed to which people adopt new practices, new behaviors and new habits. For people to adopt the change they must first choose to do so. We use the word adopt because we are talking about employees accepting change, assuming ownership for the change, and ultimately, implementing the change. Your role as a leader in the change process is to get people to choose change. For people to choose change they must want change, they must have a desire to do something different then the way it's done today. They must have conviction that somehow the change will bring value to them <u>personally</u>.

As adults, value is driven from our current knowledge and experiences. Value is inferred based on our beliefs about a given situation, which are the result of conclusions drawn from the meaning assumed through observations and data. In order to inspire individuals to choose change you, as a change leader, must create new knowledge and experiences and provide an opportunity for individuals to change their beliefs. A change in belief, however, only comes as a result of quantifying the value gained through our own observations within a specific experience. Therefore, the new experience must be relevant and meaningful to the individual or group of individuals that you are expecting to see a change in behavior. We can simplify the model for individual change using the following formula:

$$K_n + E_n + V = \Delta$$

Where new knowledge (K_n) plus a new experience (E_n) and the value (V) perceived from the experience creates the ability for an individual to choose change.

Let's look at a practical example of this formula for change from a popular reality show, "The Biggest Loser". On this reality show, contestants are selected to participate in an opportunity to transform their lives through fitness and a healthy diet. Each group of contestants has a professional trainer, as their change leader, to guide them and motivate them through the transformation. If you've watched this show you know that change only comes when the contestant makes a personal decision to go beyond the pain, beyond the overwhelming emotions, and adopt a new lifestyle. Those that don't make this decision don't lose the weight and are voted off the show.

On one particular episode, the trainers brought in an expert to show the contestants how to cook healthy meals using everyday ingredients. In this example, the change leaders are providing new knowledge and creating a new experience. The expert demonstrates how to cook a loaded sweet potato, providing healthy alternatives like Greek yogurt for unhealthy ingredients like sour cream in order to reduce calories while maintaining flavor. Recognizing that some were skeptical of the yogurt on a potato, they portioned out the new meal so each contestant would have the opportunity to taste and examine the new ingredients together for themselves. The trainers then, in their small groups, provided their team with an opportunity to express their observations. This starts the value discussion immediately, and a couple participants even begin to communicate conclusions from their experience, saying "I never believed that yogurt and potatoes would be so good together" and "I'll definitely make this for my family". With the new knowledge and experience, each contestant was allowed to form their own beliefs, and given the ability choose change.

Now let's put this practice into play at your facility, within your improvement initiative. The first thing you need to do is identify the group of people that need to change, and the new behaviors you expect to see as a result. Then, determine what new knowledge must be shared with the group to help them make a judgment about the change. The next step is for you, the change leader. Decide how best to guide the group through a new experience during which they can make observations and gather new data to form their own beliefs about the change. Finally, you'll need to define the reinforcing system that will encourage the new expected behaviors and ensure old habits don't resurface. Use the example provided on the following page as a visual reminder of how to build your plan for creating the ability for change.

The Six Sigma Design for Reliability

Role or Group:	Maintenance Supervisors
What are the **Current Behaviors** that must change?	• Don't wait for assets to fail, be proactive • "Spinning the bearing" is not quantitative enough to determine if the bearing is healthy • Ignoring defects identified by condition monitoring technologies
What are the expected **New Behaviors**?	• Make decisions to correct defects earlier in the P-F curve • Use quantitative measures to determine asset health and corrective actions
What **New Knowledge** must be provided?	• Asset Health Management • P-F Management Philosophy • Proactive vs. Reactive Decision Making
How will you create a **New Experience** to enable new beliefs and inferred value?	
Leadership Decision	• Pick 3 components to replace within the next 30 days that have an identified bearing defect
Scheduled Event	• In collaboration with tradesmen, examine the replaced component to confirm defect identified by condition monitoring technology
Observations & Conclusions *(captured during the event)*	• *Observed that the surface fatigue discoloration on the main drive bearing was not in the center of the race* • *Concluded that the bearing was not installed properly* • *Observed that the main drive bearing was not heavily deteriorated like they expected* • *Concluded that the technology did identify a defect long before any visual or audio evidence existed* • *Concluded that finding the bearing defect sooner made it easier to identify the cause*
Inferred Value *(captured after the event)*	• *"Condition monitoring technologies saves me time when troubleshooting" – Fred the Mechanic* • *Making decisions sooner reduces the time and cost of the repair*
How will you **Reinforce** the new and extinguish the old behaviors?	• At weekly all-hands meetings, praise those work centers who resolve >80% of the identified defects when work order priority is a 3 or 4 • Personally follow up with work center supervisors who resolve <60% of the identified defects when work order priority is a 3 or 4 and determine course of action to improve performance

Auditing for Compliance

There are three primary factors that can drive an improvement project away from the desired outcomes:

1. The rate at which your stakeholders let go of the old way and embrace the new standards of practice,
2. How well people apply new processes and practices in relation to the expected results, and
3. Participation within the change, and the level of compliance to new processes and practices.

Some may say that compliance, the third factor, is a result of both the rate of adoption and proficiency of application, the first two factors. We might also say that without an ability to change, as we have already discussed, compliance, or the expectation of compliance is unrealistic. Before we can begin to measure the level by which our stakeholders are complying with the new practices, we must first establish the ability to change.

When developing your audit forms and methods, it is important to balance your "test" of the new solutions across four key elements of compliance:

1. Compliance to Implementation – to test and verify that the new processes and practices are working in accordance with design.
2. Compliance to Performance – to test and verify that the new processes are yielding the expected results.
3. Compliance in Execution – to test the level of integration into daily routines as observed through the operation of each process.
4. Compliance by the Numbers – to test daily management systems and the associated performance indicators in order to ensure that the impact of your solutions is recognizable.

Finally, "control" implies that you are able to recognize when the intended solution is not going to yield the desired result. Take for example a race car driver, who, with the help of his Engineer, is deploying a solution that gives him an edge over other cars on the track. The solution? A new line around the track, one that is matched perfectly, through analysis, with the setup of his race car. How does the Driver know that 1) he's on the best line, and 2) that he's going to win the race?

In this example, the Engineer or Crew Chief is measuring the Driver's lap times, and providing immediate feedback to the Driver so he can adjust his line around the track with each lap.

The same is true for your improvement process. Like the Crew Chief or Engineer, you must monitor the solution as it is being implemented, noting the variability in results, and coaching those who are executing the solution, like the Driver, to make minor adjustments to the solution in order to achieve the desired outcome. This can be accomplished using key performance indicators – a metric that is measured against a specific objective or outcome – and routine audits as we have already explained. Key performance indicators should be monitored daily during the implementation, and audits should be conducted randomly over the course of the first 9 weeks after implementation until which time you are confident that a "Compliance to Implementation" risk is unlikely.

After the first 9 weeks, audits transform and are managed within the natural Daily Management System. Audits are conducted when a key performance indicator, within the management system, failed to meet the expected result. Again, these indicators should help us forecast the way forward, not look in the past. The purpose of the audit within the second 9 weeks is to evaluate the "Compliance in Execution" risks, looking specifically for evidence of work-arounds or old habits that will ultimately derail the solutions if left unchecked.

VOLUME SUMMARY

The *Six Sigma Design for Reliability*, like many Six Sigma continuous improvement methodologies, begins with a clear definition of the problem needing to be solved. In terms of Asset Management and the business of reliability or maintenance, improvement begins by selecting the business objectives.

Before committing to an improvement path, take the time needed to measure the current process and associated performance variables that are preventing your organization from achieving the desired objectives. With a firm understanding of priorities, analyze the root causes of performance variation and select solutions that will lend themselves to driving performance improvement in 26 weeks (6 months) or less.

Don't lose site of the fact that the improvement process will impact people, and relies on people to make the dream a reality. Construct your "Should-Be" process with this in mind. Distribute role responsibilities within the new process, and create the ability for people to adopt the new process within their day-to-day routines.

What's Next?

In the next several volumes of the *Six Sigma Design for Reliability*, we will dive deeper into the solutions. We will examine best practices associated with the Maintenance Work Management process, and explore the methods associated with Reliability Centered Maintenance, including Failure Modes & Effects Analysis, Equipment Maintenance Planning, and how to select the right Condition Monitoring Technologies for your improvement process.

In a subsequent volume we will also take a tour through Reliability Analytics, including Reliability Modeling, Process Capability Analysis, and Hypothesis testing in order to refine our understanding of statistics that can be used during both the Measure and Control phases of the *Six Sigma Design for Reliability* improvement process.

ADDITIONAL RESOURCES

Key Performance Indicators

On-time In-full Delivery (Goods & Services)	99%
Recordable EH&S Events (per 200k Hours)	0.5
Employee Turnover (per Year)	3.0%
Absentees (per Year)	1.0%
Training Budget (per Employee / per Year)	$1,350
Monthly Budget Variance (Overall)	< 2%
Overall Equipment Effectiveness	85%
Operations Availability (Uptime)	95%
Operations Efficiency (Rate)	95%
Single-Pass Quality	95%
Defect Rate per Million Opportunities (dpm)	100
Waste or Scrap Costs (% Manufacturing Costs)	0.2%
Maintenance Spending (% RAV)	< 4.0%
Maintenance Craftsmen per $7M RAV	1
Contract Service Costs (% Maintenance Spending)	5.0%
Overtime (% Straight-Time)	5%
Direct-Labor Utilization Rate	65%
Schedule Compliance	90%
Unscheduled Maintenance (% Total Maintenance)	< 10%
Preventive/Predictive Maintenance (% Total Maintenance)	30%
Schedule Efficiency	85%
Maintenance Backlog (Maximum Total Weeks)	6
Maintenance Backlog (Maximum Ready to Schedule)	4
Total Inventory Value (% RAV)	1.0%

Warehouse Personnel per $1M TIV	1
Inventory Carrying Costs (% TIV)	< 24%
Expedited Purchases (% Total Purchases)	< 2.0%
Stock-Outs (per Month)	< 1
Inventory Turns (Average per Year)	2.0
Inventory Accuracy	98%
On-Time Deliveries	98%
Warehouse Space Utilization	85%
Warehouse Service Level	95%
Maximum Inventory Days-On-Hand	28

Business Finance Terms and Definitions

Gross Margin

Gross margin is a company's total sales revenue minus its cost of goods sold (COGS), divided by total sales revenue, expressed as a percentage. The gross margin represents the percent of total sales revenue that the company retains after incurring the direct costs associated with producing the goods and services it sells. The higher the percentage the more cash the company retains on each dollar of sales to sustain business operations.

$$Gross\ Margin = \frac{Revenue - COGS}{Revenue} \quad \text{or} \quad Gross\ Margin = \frac{Gross\ Income}{Sales}$$

A declining gross margin may indicate that the cost of goods sold, or variable costs per unit produced are increasing, or that revenues are decreasing without a proportional decrease in variable costs. A disproportionate variable cost per volume of sales is often the result of maintaining direct material inventories above the levels required to meet production demand.

Delays in accounts receivable – the money owed to the company by its customers – may also temporarily reduce gross margin if the company pays for direct materials or services – accounts payable – before revenues are recognized.

Profit Margin

Profit margin is net income divided by revenue, or net profits divided by sales. Net income or net profit may be determined by subtracting all of a company's expenses, including operating costs, variable costs (e.g. Cost of Goods Sold) and tax costs, from its total gross revenue. Profit margins are expressed as a percentage and, in effect, measure how much out of every dollar of sales a company actually retains as earnings or "equity". A 20% profit margin, then, means the company has a net income of $0.20 for each dollar of total revenue earned.

$$Profit\ Margin = \frac{Net\ Income}{Revenue}$$

Profit margin also indicates a company's ability to manage its expenses. High expenditures relative to revenue (i.e. a low profit margin) may indicate that a company is struggling to manage its fixed operating costs, perhaps because of an increase in overhead costs, depreciation or maintenance costs. High expenditures may occur for many reasons, including that the company has too much inventory relative to its sales, that it has too many employees relative to sales, or that it is operating in spaces that are too large and thus the company is paying too much in rent or utilities.

Income Statement

An income statement is a financial statement that reports a company's financial performance over a specific accounting period. Financial performance is assessed by giving a summary of how the business incurs its revenues and expenses through both operating and non-operating activities. It also shows the net profit or loss incurred over a specific accounting period.

$$Net\ Income = Revenue - Costs$$

Income Statement

($1,000's)

Account	2015	2014	2013
Sales (Revenue)	5,800	5,156	5,099
Cost of Goods Sold (COGS)	3,062	3,642	2,885
Gross Income	2,738	1,514	2,214
Sales, General and Administrative (SG&A)	1,369	757	1,107
Operating Income	1,369	757	1,107
Maintenance Expense	479	265	387
Depreciation Expense	137	76	111
Interest Expense	68	38	55
Income Before Taxes	685	379	554
Income Tax Expense	185	102	149
Net Income	500	276	404

Sales, General and Administrative (SG&A) Costs

Sales expenses include shipping supplies, delivery charges and sales commissions, as well as, indirect expenses that occur throughout the manufacturing process and after the product is finished. An item does not have to be sold for an indirect expense to be incurred. Indirect expenses include product advertising and marketing, telephone bills, travel costs and the salaries of sales personnel.

General and Administrative (G&A) expenses are referred to as the overhead of the company. They are the costs a company must incur to open the doors each day. G&A costs are more fixed than selling costs because they include rent/mortgage on buildings, utilities and insurance. G&A costs also include salaries of all non-sales "operations" personnel.

The Six Sigma Design for Reliability

Balance Sheet

A balance sheet is a financial statement that summarizes a company's assets, liabilities and shareholders' equity at a specific point in time. These three balance sheet segments give investors an idea of what the company owns and owes, as well as the amount invested by shareholders, since the company began.

$Assets = Liabilities + Equity$

Balance Sheet

($1,000's)

Account	2015	2014	2013
Assets			
Current Assets			
Cash	276	404	313
Accounts Receivable	580	515	509
Inventory	1,531	1,802	1,400
Total Current Assets	2,387	2,721	2,222
Property and Equipment	3,992	2,650	2,419
Total Assets	6,379	5,371	4,641
Liabilities			
Current Liabilities			
Notes (Loans) Payable	644	1,170	1,560
Accounts Payable	383	451	350
Income Taxes Payable	135	75	109
Total Current Liabilities	1,162	1,695	2,019
Long-term Debt	2,783	1,441	1,209
Total Liabilities	3,944	3,136	3,228
Equity			
Common Stock	24	22	14
Retained Earnings	2,410	2,213	1,398
Total Equity	2,435	2,235	1,412
Total Liabilities and Equity	6,379	5,371	4,641

ASSETS, in financial terms, are the resources owned by the company, such as cash, inventory, land, property and equipment that can be used to generate revenue. LIABILITIES are monies owed by the company to lenders, suppliers, shareholders and even the government by way of taxes. EQUITY in a business is the capital contributed by investors or owners, plus the retained earnings of the business for future reinvestment in the company's growth or development as "working capital".

Let's demonstrate these definitions using a home mortgage. If the initial purchase of the home is $100,000 then your total ASSETS equal $100,000. If you borrowed $80,000 using a home mortgage loan, then your total LIABILITY is $80,000 and your EQUITY in the home is $20,000 – the amount of money invested by you, the owner, in the home. As you pay down the mortgage loan your liabilities will decrease and your equity will increase, but your total assets will remain $100,000 – the value if sold by you, the asset owner.

Asset Efficiency

Asset efficiency is a financial ratio that measures the turnover of a company's assets while generating sales revenue or sales income for the company.

$$Asset\ Efficiency = \frac{Revenue}{Average\ Total\ Assets}$$

As an example, comparing the asset turnover ratios for AT&T Inc. (T) and Verizon Communications Inc. (VZ) provides us with one industry's picture of asset efficiency.

AT&T Inc. (T) had total revenues of $132 billion when the fiscal year ended on December 31, 2014. Total assets at the beginning and end of the 2014 fiscal year were $278 billion and $293 billion respectively, for an average asset base of $287 billion. AT&T's asset turnover ratio in 2014 was therefore 46% ($132 billion / $287 billion).

Verizon had total revenues of $127 billion. Total assets at the beginning and end of the year were $274 billion and $232 billion, respectively, for an average asset base of $253 billion. As such, in 2014 Verizon's asset turnover ratio was 50% ($127 billion / $253 billion), about 9% higher than AT&T's in the same year despite having a lower annual revenue.

(Source: Investopedia)

Return On Assets (ROA) and Return On Equity (ROE)

RETURN ON ASSETS (ROA) is an indicator of how profitable a company is relative to its total assets. ROA tells us how efficient management is at using its assets to generate earnings. Calculated by dividing a company's annual Net Income – earnings or profit – by its total assets, ROA is displayed as a percentage. Sometimes this is referred to as "return on investment".

$$ROA = \frac{Net\ Income}{Average\ Total\ Assets}$$

RETURN ON EQUITY (ROE) is the amount of Net Income returned as a percentage of equity. ROE measures a company's value per share by revealing how much profit a company generates with the money shareholders have invested.

$$ROE = \frac{Net\ Income}{Average\ Total\ Equity}$$

RETURN ON ASSETS and RETURN ON EQUITY are metrics used to assess the financial performance of a business relative to the resources it uses, like capital and cash, to generate income or profits within a fiscal period, such as a month, quarter or calendar year.

The numerator in both the RETURN ON ASSETS and RETURN ON EQUITY metrics is "Net Income". Net Income is the sum of Sales Revenue minus the Cost of Goods Sold, SG&A, and other expenses incurred by the business within the same period of time as Sales.

As an example, if we assume the company sold $1,000 worth of a product, and it cost the company $400 to make the product, and another $300 of operating, sales and general administrative costs, the Net Income would be $300. If we then divide this $300 by an average total assets, for the same period of time, valued at $10,000, then our RETURN ON ASSETS is 0.03, or 3%.

If we assume that Liabilities total $4,000 and the total Equity is $6,000, then dividing this $300 by the average total equity we arrive at a RETURN ON EQUITY of 0.05, or 5%.

When comparing these two financial ratios, RETURN ON EQUITY will always be a larger ratio or percentage because RETURN ON ASSETS includes the business liabilities and evaluates Net Income, or the company's ability to generate profits, based on the total Assets. RETURN ON EQUITY, on the other hand, only evaluates Net Income based on contributed capital and retained earnings or "working capital".

Business Case Calculations Cheat Sheet

Assets

$Assets = Liabilities + Equity$

Availability

$$A(t) = \frac{R(t)}{R(t) + M(t)} \quad \text{or} \quad A = \frac{Scheduled\ Capability}{Theoretical\ Capability}$$

Asset Utilization

$AU = Availability * Rate * Quality$

Backlog

The Six Sigma Design for Reliability

$$Backlog = \frac{Total\ Estimated\ Work\ Order\ Labor\ Hours}{Net\ Available\ Labor\ Hours\ per\ Week}$$

Consequence *Effect*

$$Effect = \frac{Net\ Income}{Time}$$

Maintainability (Probability)

$$M(t) = 1 - \left(e^{-\left(\frac{Time}{MTTR}\right)}\right)$$

Net Available Labor

$$NAL = (Payroll\ Hours - Indirect\ Hours)$$

Net Income

$$Net\ Income = Revenue - (COGS + Expenses)$$

Net Present Value

$$NPV = \sum_{t=1}^{Payback\ Period} \frac{Net\ Income}{(1 + Discount\ Rate)^t}$$

Overall Equipment Effectiveness

$$OEE = Uptime * Rate * Quality$$

Reliability (Probability)

$$R(t) = e^{-(\frac{Time}{MTBF})}$$

Replacement Asset Value

$$RAV = \frac{Annual\ Maintenance\ Costs}{Asset\ Replacement\ Cost}$$

Return On Assets (ROA)

$$ROA = \frac{Net\ Income}{Average\ Total\ Assets}$$

Return On Equity (ROE)

$$ROE = \frac{Net\ Income}{Average\ Total\ Equity}$$

Risk

$$Risk = Probability * Consequence\ Effect$$

Weibull Distribution

The Six Sigma Design for Reliability

$$R(t) = e^{-(\frac{Time}{\eta})^{\wedge}\beta}$$

www.ingramcontent.com/pod-product-compliance
Lightning Source LLC
Chambersburg PA
CBHW061447180526
45170CB00004B/1600